ENERGY SECTOR STANDARD OF THE PEOPLE'S REPUBLIC OF CHINA

中华人民共和国能源行业标准

Code for Design of Gas-Insulated Metal-Enclosed Switchgear Installations

气体绝缘金属封闭开关设备配电装置设计规范

NB/T 35108-2018

Replace DL/T 5139-2001

Chief Development Department: China Renewable Energy Engineering Institute

Approval Department: National Energy Administration of the People' s Republic of China

Implementation Date: July 1, 2018

China Water & Power Press

中国水利水电出版社

Beijing 2024

图书在版编目（CIP）数据
气体绝缘金属封闭开关设备配电装置设计规范 : NB/T 35108-2018 代替 DL/T 5139-2001 = Code for Design of Gas-Insulated Metal-Enclosed Switchgear Installations（NB/T 35108-2018 Replace DL/T 5139-2001） : 英文 / 国家能源局发布. -- 北京 : 中国水利水电出版社, 2024. 7. -- ISBN 978-7-5226-2658-1
Ⅰ. TM564-65
中国国家版本馆CIP数据核字第2024LS7947号

ENERGY SECTOR STANDARD
OF THE PEOPLE'S REPUBLIC OF CHINA
中华人民共和国能源行业标准

Code for Design of Gas-Insulated
Metal-Enclosed Switchgear Installations
气体绝缘金属封闭开关设备配电装置设计规范

NB/T 35108-2018
Replace DL/T 5139-2001
（英文版）

Issued by National Energy Administration of the People's Republic of China
国家能源局　发布
Translation organized by China Renewable Energy Engineering Institute
水电水利规划设计总院　组织翻译
Published by China Water & Power Press
中国水利水电出版社　出版发行
Tel: (+ 86 10) 68545888　68545874
sales@mwr.gov.cn
Account name: China Water & Power Press
Address: No.1, Yuyuantan Nanlu, Haidian District, Beijing 100038, China
http: //www.waterpub.com.cn
中国水利水电出版社微机排版中心　排版
北京中献拓方科技发展有限公司　印刷
184mm×260mm　16 开本　3 印张　95 千字
2024 年 7 月第 1 版　2024 年 7 月第 1 次印刷
Price（定价）: ¥ 495.00

Introduction

This English version is one of China's energy sector standard series in English. Its translation was organized by China Renewable Energy Engineering Institute authorized by National Energy Administration of the People's Republic of China in compliance with relevant procedures and stipulations. This English version was issued by National Energy Administration of the People's Republic of China in Announcement [2023] No. 1 dated February 6, 2023.

This version was translated from the Chinese Standard NB/T 35108-2018, *Code for Design of Gas-Insulated Metal-Enclosed Switchgear Installations,* published by China Water & Power Press. The copyright is reserved by National Energy Administration of the People's Republic of China. In the event of any discrepancy in the implementation, the Chinese version shall prevail.

Many thanks go to the staff from the relevant standard development organizations and those who have provided generous assistance in the translation and review process.

For further improvement of the English version, any comments and suggestions are welcome and should be addressed to:

China Renewable Energy Engineering Institute
No. 2 Beixiaojie, Liupukang, Xicheng District, Beijing 100120, China
Website: www.creei.cn

Translating organizations:

Changjiang Survey, Planning, Design and Research Co., Ltd.

CSG Power Generation Co., Ltd.

Translating staff:

XIONG Weijun LIANG Bo CUI Lei CHEN Changxu
HUANG Wei CHENG Zhuang WANG Shuqing XIAO Jun
LI Dehua ZHU Zhao JIAN Wei LIU Yaqing
MA Lingteng ZHU Zheng RONG Xuening LI Mengbo
XIONG Manni CHEN Gong XU Zecheng CHEN Baisen
LU Xiaojun LE Lingling

Review panel members:

GUO Jie POWERCHINA Beijing Engineering Corporation Limited

GAO Yan	POWERCHINA Beijing Engineering Corporation Limited
CHEN Gang	POWERCHINA Huadong Engineering Corporation Limited
LI Zhongjie	POWERCHINA Northwest Engineering Corporation Limited
CHEN Lei	POWERCHINA Zhongnan Engineering Corporation Limited
LI Qian	POWERCHINA Zhongnan Engineering Corporation Limited
HOU Yujing	China Institute of Water Resources and Hydropower Research
ZHANG Ming	Tsinghua University

National Energy Administration of the People's Republic of China

翻译出版说明

本译本为国家能源局委托水电水利规划设计总院按照有关程序和规定，统一组织翻译的能源行业标准英文版系列译本之一。2023年2月6日，国家能源局以2023年第1号公告予以公布。

本译本是根据中国水利水电出版社出版的《气体绝缘金属封闭开关设备配电装置设计规范》NB/T 35108—2018 翻译的，著作权归国家能源局所有。在使用过程中，如出现异议，以中文版为准。

本译本在翻译和审核过程中，本标准编制单位及编制组有关成员给予了积极协助。

为不断提高本译本的质量，欢迎使用者提出意见和建议，并反馈给水电水利规划设计总院。

地址：北京市西城区六铺炕北小街2号
邮编：100120
网址：www.creei.cn

本译本翻译单位：长江勘测规划设计研究有限责任公司
南方电网调峰调频发电有限公司

本译本翻译人员：熊为军 梁 波 崔 磊 陈昌旭
黄 炜 程 壮 王树清 肖 军
李德华 朱 钊 简 巍 刘亚青
马凌腾 朱 正 荣雪宁 李梦柏
熊曼妮 陈 功 徐则诚 陈柏森
鲁晓军 乐零陵

本译本审核人员：

郭 洁 中国电建集团北京勘测设计研究院有限公司
高 燕 中国电建集团北京勘测设计研究院有限公司
陈 钢 中国电建集团华东勘测设计研究院有限公司
李仲杰 中国电建集团西北勘测设计研究院有限公司
陈 蕾 中国电建集团中南勘测设计研究院有限公司
李 倩 中国电建集团中南勘测设计研究院有限公司

侯瑜京　中国水利水电科学研究院
张　明　清华大学

国家能源局

Announcement of National Energy Administration of the People's Republic of China [2018] No. 4

According to the requirements of Document GNJKJ [2009] No. 52, "Notice on Releasing the Energy Sector Standardization Administration Regulations (*tentative*) and detailed implementation rules issued by National Energy Administration of the People's Republic of China", 168 sector standards such as *Guide for Evaluation of Vibration Condition for Wind Turbines*, including 56 energy standards (NB) and 112 electric power standards (DL), are issued by National Energy Administration of the People's Republic of China after due review and approval.

Attachment: Directory of Sector Standards

National Energy Administration of the People's Republic of China
April 3, 2018

Attachment:

Directory of Sector Standards

Serial number	Standard No.	Title	Replaced standard No.	Adopted international standard No.	Approval date	Implementation date
…						
28	NB/T 35108-2018	Code for Design of Gas-Insulated Metal-Enclosed Switchgear Installations	DL/T 5139-2001		2018-04-03	2018-07-01
…						

Foreword

According to the requirements of Document GNKJ [2012] No. 83 issued by National Energy Administration of the People's Republic of China, "Notice on Releasing the Development and Revision Plan of the First Batch of Energy Sector Standards in 2012", and after extensive investigation and research, summarization of practical experience, and wide solicitation of opinions, the drafting group has prepared this code.

The main technical contents of this code include: basic requirements, configuration and interface, type selection of components, layout, environment protection, earthing, requirements for civil design, allocation of special tools and instruments, and site test.

The main technical contents revised are as follows:

— Incorporating "Scope" into "General Provisions".

— Incorporating "General Requirements" into "Basic Requirements" and "Configuration and Interface" respectively.

— Changing the chapter "GIS Distribution Device Selection" into "Component Selection".

— Adding the content of "Layout of Site Withstand Voltage Test Equipment" and "Expansion".

National Energy Administration of the People's Republic of China is in charge of the administration of this code. China Renewable Energy Engineering Institute has proposed this code and is responsible for its routine management. Energy Sector Standardization Technical Committee on Hydropower Electrical Design is responsible for the explanation of specific technical content. Comments and suggestions in the implementation of this code should be addressed to:

China Renewable Energy Engineering Institute
No. 2 Beixiaojie, Liupukang, Xicheng District, Beijing 100120, China

Chief development organizations:

Changjiang Survey, Planning, Design and Research Co., Ltd.

Electrical Professional Committee of China Society for Hydropower Engineering

Chief drafting staff:

SHI Fengxiang YUAN Ailing CHEN Changbin CHEN Xiaoming,
WANG Huajun MAO Yongsong LI Dingzhong LIANG Bo

CHENG Zhuang CAI Bin CUI Lei CHEN Changxu

JIAN Wei ZHU Zhao

Review panel members:

YU Qinggui WANG Runling FANG Hui KANG Benxian

FENG Zhenqiu WANG Yaohui SHAO Guangming YANG Jianjun

XU Lijia WANG Xiaobing WANG Yong LI Yong

XIA Fujun YANG Yuhu SANG Zhiqiang YANG Yunfeng

Contents

1 General Provisions

1.0.1 This code is formulated with a view to standardizing the design of gas-insulated metal-enclosed switchgear installations, to achieve the objectives of technological advancement, safety, reliability, economy, quality, and easy maintenance.

1.0.2 This code is applicable to the design of gas-insulated metal-enclosed switchgear installations with a nominal voltage of 66 kV to 750 kV and a frequency of 50 Hz for hydropower station and substation.

1.0.3 In addition to this code, the design of gas-insulated metal-enclosed switchgear installations shall comply with other current relevant standards of China.

2 Terms

2.0.1 gas-insulated metal-enclosed switchgear (GIS)

metal-enclosed switchgear in which the insulation is obtained, at least partly, by an insulating gas with a pressure higher than air atmospheric pressure, GIS for short

2.0.2 transport unit

part of GIS suitable for transportation without being disassembled

2.0.3 enclosure

component of GIS, which is used to accommodate insulating gas and bear the specified pressure, to keep the specified insulation level, protect the equipment against external influences, and provide safety protection for personnel under specified conditions

2.0.4 compartment

part of GIS, which is enclosed completely except for openings required for interconnection and control

2.0.5 component

essential part of the main or earthing circuits of GIS which undertakes a specific function

2.0.6 support insulator

internal insulator supporting one-phase or multi-phase conductor

2.0.7 partition

support insulator of GIS for separating two adjacent compartments

2.0.8 bay

spatial structure of a bay, which is characterized by the width and layout of main components. A bay generally consists of one functional unit, and is sometimes called the incoming bay, outgoing bay, etc. according to the unit function

2.0.9 buffer junction

connecting part between two adjacent enclosures, which is used to adjust installation error, compensate the displacement caused by the thermal expansion and contraction of enclosure and the uneven settlement of foundation

2.0.10 pipeline busbar

part of GIS, including main busbar, branch busbar and the connecting pipeline between GIS components

2.0.11 main circuit

all the conductive parts of GIS which is used to transmit electrical energy

2.0.12 auxiliary circuit

all the conductive parts included in the control, measurement, signal and regulation circuits of GIS

2.0.13 gas-insulated metal-enclosed transmission line (GIL)

metal-enclosed transmission line in which the insulation is obtained, at least partly, by an insulating gas with a pressure higher than air atmospheric pressure, GIL for short

3 Basic Requirements

3.1 Normal Service Conditions

3.1.1 GIS is classified into indoor and outdoor types. If the bulk of the GIS is installed indoors and only the ingoing and outgoing equipment is installed outdoors, except for the outdoor part that shall meet the outdoor service conditions, other parts shall be designed according to the indoor service conditions.

3.1.2 The indoor switchgear installation shall be designed according to the following environmental conditions:

1 The maximum temperature of the ambient air does not exceed 40 °C and the average temperature measured within 24 h does not exceed 35 °C. The preferred values of the minimum temperature of ambient air are −5 °C, −15 °C, and −25 °C.

2 The altitude does not exceed 1000 m above sea level.

3 For the air humidity, the possible condensation during the high humidity period shall be considered according to the climatic conditions in different regions. The daily average relative humidity does not exceed 95 % and the monthly average does not exceed 90 %.

4 The ambient air shall be free of significant pollution by dust, smoke, corrosive gas, combustible gas, vapor, and salt fog.

3.1.3 The outdoor switchgear installation shall be designed according to the following environmental conditions:

1 The maximum temperature of the ambient air does not exceed 40 °C and the average temperature measured within 24 h does not exceed 35 °C. The preferred values of the minimum ambient air temperature are −10 °C, −25 °C, and −30 °C. The preferred values of the daily temperature difference are 15 K and 25 K.

2 The daily average relative humidity is 100 % when considering condensation and precipitation.

3 The altitude does not exceed 1000 m above sea level.

4 The wind speed does not exceed 34 m/s, which is equivalent to that the wind pressure on a cylinder surface does not exceed 700 Pa.

5 The preferred thicknesses of ice coating are 1 mm, 10 mm, and 20 mm.

6 The intensity of solar radiation is 1000 W/m^2, while the ambient air temperature is 40 °C and the wind speed is 0.5 m/s.

7 The pollution level of air does not exceed Level d as defined in the current national standard GB/T 26218.1, *Selection and Dimensioning of High-Voltage Insulators Intended for Use in Polluted Conditions—Part 1: Definitions, Information and General Principles.*

8 According to the climatic conditions in different regions, the influences of hail and sudden temperature change shall be considered.

3.2 Special Service Conditions

3.2.1 When GIS is used under the conditions different from the normal service conditions given in Section 3.1, corrections shall be made in accordance with the current national standard GB/T 11022, *Common Specifications for High-Voltage Switchgear and Controlgear Standards* and the requirements shall be put forward for particular environmental conditions.

3.2.2 When the altitude is higher than 1000 m above sea level, corrections shall be made to the rated insulation level of external insulation equipment, the safety clearance of switchgear installation, and the pressure meters.

3.2.3 The seismic design of GIS shall comply with the current national standards GB/T 13540, *Seismic Qualification for High-Voltage Switchgear and Controlgear* and GB 50260, *Code for Seismic Design of Electrical Installations.*

3.2.4 For equipment used in seriously polluted air, the pollution level shall conform to Level e as defined by the current national standard GB/T 26218.1, *Selection and Dimensioning of High-Voltage Insulators Intended for Use in Polluted Conditions—Part 1: Definitions, Information and General Principles.*

3.2.5 The preferred minimum temperature of ambient air in extremely cold climate should be −40 °C and −50 °C. The preferred maximum temperature of ambient air in extremely hot climate should be 45 °C, 50 °C and 55 °C.

3.2.6 The relative humidity shall be determined by negotiation between the manufacturer and the user when the average relative humidity measured within 24 h reaches 98 % indoors in humid tropics.

3.2.7 The ice coating thickness shall be determined by negotiation between the manufacturer and the user when the ice thickness exceeds 20 mm.

3.2.8 The wind speed shall be determined by negotiation between the manufacturer and the user when the design wind speed exceeds 34 m/s.

3.3 Performance Parameters

3.3.1 The design of GIS shall consider both short-term and long-term use, according to engineering characteristics, scale and development planning.

3.3.2 The design of GIS shall follow the principle of saving land.

3.3.3 The selection of GIS parameters shall meet the requirements of normal operation, maintenance, short circuit, overvoltage, etc., and shall meet the long-term development.

3.3.4 For the ambient air temperature for outdoor GIS, the average annual maximum temperature or the average annual minimum temperature shall be selected.

3.3.5 For the relative humidity of outdoor GIS, the average annual relative humidity of the highest humidity month shall be selected. In humid tropics, electrical equipment of humid tropics type shall be used. In sub-humid tropics, ordinary electrical equipment may also be used, but measures against moisture, condensation, water, rust, mildew, and insect pests shall be strengthened according to local practice.

3.3.6 When the ambient air temperature is lower than the minimum allowable temperature of the equipment, instruments and relays of GIS, heating devices or other thermal insulation facilities shall be installed.

3.3.7 For outdoor GIS of 330 kV and below, the maximum design wind speed may adopt the 10-min average maximum wind speed of a 30-year return period at 10 m above the ground. For outdoor GIS of 500 kV and above, the maximum design wind speed should adopt the 10-min average maximum wind speed of a 50-year return period at 10 m above the ground. When the maximum design wind speed exceeds 34 m/s, the fixing between GIS and foundation shall be strengthened.

3.3.8 The layout of GIS should avoid the vibration area with a continuous vibration source, otherwise, anti-vibration measures shall be taken.

3.3.9 GIS shall meet the requirements of various possible operating modes; the maximum allowable operating voltage shall not be lower than the rated voltage of the circuit, and the long-term allowable current shall not be lower than the maximum continuous operating current of the circuit. The voltage level shall comply with the current national standard GB/T 156, *Standard Voltages*. The current value shall comply with the current national standard GB/T 762, *Standard Current Ratings*.

3.3.10 The insulation level of GIS shall be determined based on the various overvoltage appearing on GIS and the characteristics of protection devices. The costs of the various overvoltage protection devices, comprehensive investment, maintenance costs and losses due to failure shall be taken into account in the insulation coordination, to achieve the best comprehensive economic benefits.

3.3.11 The rated peak withstand current, rated short-time withstand current and rated short-circuit breaking current of GIS shall be calculated according to the planned capacity of the project, considering the long-term development plan of the electric power system for 5 years to 10 years after putting into operation. Rated short-circuit current shall be determined according to the normal wiring mode of the maximum short-circuit current that might occur, and the influence of DC component of short-circuit current shall also be considered. If the development of the power system is not clear, the selection of short-circuit current shall comply with the provisions on short-circuit current level control year in the current sector standard of China DL/T 5429, *Technical Code of Design for the Electric Power System*.

3.3.12 Temperature rise of each component of GIS shall comply with the current national standard GB/T 11022, *Common Specifications for High-Voltage Switchgear and Controlgear Standards*. The allowable temperature rise of enclosure shall be in accordance with Table 3.3.12. The parts with a temperature rise exceeding 40 K shall be marked with obvious signs of high temperature to warn the maintenance personnel. Also, the surrounding insulating materials and sealing materials shall be ensured not to be damaged.

Table 3.3.12 Allowable temperature rise of enclosure (K)

Enclosure part	**Allowable temperature rise at the ambient air temperature 40 °C**
Easily touched by operator	≤ 30
Easily touched but no need to be touched during operation by operator	≤ 40
Not easily touched by operator	≤ 65

3.3.13 GIS enclosure shall be made of non-magnetic metal materials. Aluminum alloy enclosure shall comply with the current national standard GB/T 28819, *Aluminum Alloy Enclosures for Gas-Filled High-Voltage Switchgear*.

3.3.14 If GIS is to be built by stages or expanded, the disconnector and compartments should be set on the connection interface of GIS. The disconnector compartment shall be able to withstand the rated insulation level across the isolating distance.

3.3.15 For the GIS of 220 kV and below, the busbar surge arresters and voltage transformers should not be equipped with disconnector but with breaks or independent compartment with detachable breaks. For the GIS of 330 kV and above, the disconnector shall not be set for the surge arresters and voltage transformers of incoming and outgoing lines and the busbar surge arresters; the busbar voltage transformers should not be equipped with disconnectors but should be equipped with breaks or independent compartment with detachable breaks.

3.3.16 The noise control design of GIS shall be taken seriously. The environmental noise limit shall comply with the current national standard GB 3096, *Environmental Quality Standard for Noise*.

3.3.17 For overhead outgoing line equipment, electrostatic induction, radio interference level and corona shall comply with the current sector standard DL/T 5352, *Technical Code for Designing High Voltage Electrical Switchgear*.

3.3.18 The allowable annual SF_6 gas leakage rate of each compartment of GIS shall not be greater than 0.5 %. The methods and criterias for qualitative and quantitative leak detection shall comply with the current sector standard DL/T 618, *Guide for Hand-Over Test of Gas-Insulated Metal-Enclosed Switchgear on Site*.

3.3.19 The protection level of the enclosure of GIS main circuit equipment, auxiliary circuit equipment and mechanical operation equipment shall comply with the current national standard GB/T 4208, *Degrees of Protection Provided by Enclosure (IP Code)* and GB/T 20138, *Degrees of Protection Provided by Enclosures for Electrical Equipment Against External Mechanical Impacts (IK Code)*.

4 Configuration and Interface

4.1 Internal Fault Protection

4.1.1 The following protection measures shall be taken for GIS internal faults at least:

1 Compartments shall be set reasonably to reduce the internal fault coverage and limit the pressure rise rate.

2 The compartment shall be provided with desiccant, gas-replenishing check valve and temperature compensating type density relay.

3 The right and reliable interlock devices shall be provided.

4 High-speed earthing switch shall be provided reasonably.

5 Overvoltage protection and insulation coordination shall comply with the current national standard GB/T 50064, *Code for Design of Overvoltage Protection and Insulation Coordination for AC Electrical Installations*.

4.1.2 If the strength of GIS enclosure is not designed based on the maximum pressure rise, the compartment shall be provided with directional pressure relief device, and field operator and equipment shall be safe when the pressure relief device operates.

4.1.3 The GIS enclosure shall be designed to withstand short-time short-circuit current. The duration and performance of enclosure to withstand the arc shall comply with the current national standard GB/T 7674, *Gas-Insulated Metal-Enclosed Switchgear for Rated Voltages of 72.5 kV and Above*.

4.2 Very Fast Transient Overvoltage

4.2.1 When GIS for a rated voltage of 330 kV and above is directly connected to the transformer or reactor, the GIS manufacturer shall calculate and analyze very fast transient overvoltage (VFTO) caused by disconnector operation.

4.2.2 The performance of GIS shall not be affected by VFTO.

4.2.3 The VFTO calculation results and analysis report for GIS shall be provided to transformer or reactor manufacturer, and then transformer or reactor manufacturer shall decide whether to reinforce the insulation of transformer or reactor windings, to eliminate the adverse effects of VFTO on transformer or reactor.

4.3 Arrangement of Earthing Switch

4.3.1 When a bay is in maintenance, its components in main electrical circuit shall be earthed. When the enclosure is opened, the main circuit shall be earthed.

4.3.2 Earthing switches shall be provided in the following locations:

1 Near the components that are connected to GIS and need to be maintained separately, such as transformer side and incoming and outgoing line side.

2 At both sides of the electrical component where main electrical circuit is disconnected, such as both sides of the circuit breaker.

3 Main busbar.

4.3.3 Type selection of earthing switch shall follow the principles below:

1 If the unenergized condition cannot be determined in advance, the fast earthing switch with the capability of making rated peak withstand current shall be provided. The earthing switch should be installed on the outgoing line side.

2 For the fast earthing switch on line side, the capacities to make and break the electromagnetic induction and electrostatic induction shall be determined by the coupling strength between multiple circuits of transmission lines on the same tower or between adjacent parallel transmission lines.

3 If the unenergized condition can be determined in advance, the maintenance earthing switch without making capacity, or with a making capacity lower than the rated peak withstand current, may be provided.

4.3.4 Measures to isolate ground potential shall be taken for part or all of earthing terminals of the earthing switch to conduct the testing and measurement of GIS.

4.3.5 When disconnector and earthing switch are integrated into a three-position switch, the switch shall be locked reliably in the disconnecting condition.

4.4 Arrangement of Surge Arrester

4.4.1 The configuration of surge arrester shall be based on the main electrical connection and protect GIS against lightning intruding overvoltage and switching overvoltage hazards in various modes of operation, and comply with the current national standard GB/T 50064, *Code for Design of Overvoltage Protection and Insulation Coordination for AC Electrical Installations*.

4.4.2 Air-insulated surge arrester should be installed at the connection of GIS to overhead line.

4.4.3 The configuration of lightning intruding overvoltage protection surge arrester for GIS of 330 kV and above shall comply with the current national standard GB/T 50064, *Code for Design of Overvoltage Protection and Insulation Coordination for AC Electrical Installations*, and should be determined through numerical simulation.

4.4.4 The configuration of lightning intruding overvoltage protection surge arrester for GIS of 220 kV and below shall comply with the current national standard GB/T 50064, *Code for Design of Overvoltage Protection and Insulation Coordination for AC Electrical Installations*.

4.4.5 Non-residual voltage or low residual voltage online detector shall be provided for surge arrester, to record the times of impact discharges and real-time leakage current of surge arrester.

4.5 Compartment Division and Accessory Arrangement

4.5.1 GIS shall be divided into several compartments, and compartment division shall meet the following requirements:

1 Equipment with different SF_6 gas pressures shall be placed in different compartments, and circuit breakers shall be placed in separate compartments.

2 Busbar compartment division shall consider the bay equipment maintenance, which should not affect the normal operation of the un-maintained bay, and shall also consider the convenience of staged installation.

3 The scope of internal faults shall be reduced, the pressure rise rate shall be limited, and the capacity and recovery time of gas recovery device shall be considered.

4 Internal faults shall be restricted within the fault compartment.

5 The voltage transformer and surge arrester should be placed in independent compartments.

6 Some components may be set as independent compartments for maintenance, and be linked with adjacent compartment for operation, such as the connections with transformer, reactor or cable.

7 Three-phase interconnected compartment shall not be set for single-phase GIS.

4.5.2 The location of the density relay and gas-replenishing check valve within the compartment shall consider the convenience of routine maintenance, and the location of the desiccant shall consider the convenience of replacement during maintenance.

4.6 Arrangement of Buffer Junction

4.6.1 The buffer junction shall be set to regulate and absorb the displacements arising from uneven settlement of foundation, civil construction error, equipment manufacturing error, installation error, compensation of temperature stress, earthquake load, temporary displacement during circuit breaker operation, and displacement caused by transformer or reactor micro-vibration.

4.6.2 Manufacturers shall propose a reasonable buffer junction configuration scheme according to the project specific conditions and the layout and structure of GIS, and shall also propose a calculation report of buffer junction configuration if necessary.

4.6.3 Buffer junction should be set at the following positions:

1 At the long busbar of GIS.

2 At the connection between busbar and bay.

3 At the position where busbar passes through civil structural joint.

4 At the connection between GIS and transformer or reactor.

4.6.4 The GIS shall meet the requirements of ±10 mm deformation in three directions caused by uneven settlement of the foundation during operation.

4.7 Operation and Control Power Supply

4.7.1 The operation and control circuit shall be powered by two independent power supplies.

4.7.2 The AC power supplies shall be three-phase 380 V or single-phase 220 V. The normal working voltage range shall be between 85 % and 110 %.

4.7.3 The DC power supplies shall be 220 V or 110 V, and the voltage amplitude shall be between −20 % and +12.5 %.

4.7.4 The control circuits of the circuit breaker closing coils shall operate reliably within the range of 80 % to 110 % of the rated voltage; the control circuits of the circuit breaker opening coils shall operate reliably within the range of 65 % to 110 % of the rated voltage; the closing or opening coils shall not operate when the control circuit voltage is equal to or less than 30 % of the DC power supply system nominal voltage.

4.7.5 When the DC control power supply is 220 V, the current of each opening or closing coil of circuit breakers shall be less than 2.5 A. When the DC control power supply is 110 V, the current of each opening or closing coil of circuit breakers shall be less than 5 A.

4.8 Signals and Signs

4.8.1 Signals to the monitoring system shall include the following:

1 The position signals of circuit breakers, disconnectors, earthing switches, etc.

2 The alarm signals of high and low oil pressure, locking signal, and low oil pressure trip signal of hydraulic mechanism.

3 Energy storage status signal of spring mechanism.

4 SF_6 gas low pressure alarm and locking signals of each compartment.

4.8.2 The alarm signals of SF_6 gas content and oxygen content in the air of GIS room shall be sent to the monitoring system.

4.8.3 Each marshalling kiosk shall have electrical primary wiring diagram of its bay. The position of partition shall be marked on the wiring diagram.

4.8.4 Partitions shall be clearly indicated.

4.9 SF_6 Gas

4.9.1 The quality, storage, operation and testing requirements of new SF_6 shall comply with the current national standards GB/T 8905, *The Guide for Management and Measuring SF_6 Gas in Electrical Equipment*, and GB/T 12022, *Industrial Sulfur Hexafluoride*.

4.9.2 During handover acceptance, SF_6 gas humidity at 20 °C in GIS shall meet the following requirements:

1 Not more than 150 μL/L for compartments with arc decomposition products.

2 Not more than 250 μL/L for compartments without arc decomposition products.

4.10 Interface

4.10.1 The direct connection between GIS and transformers or reactors shall meet the following requirements:

1 The connection between GIS and transformers or reactors shall comply with the current national standard GB/T 22382, *Direct Connection Between Power Transformers and Gas-Insulated Metal-Enclosed Switchgear for Rated Voltages of 72.5 kV and Above*.

2 GIS and transformers or reactors have independent characteristics and

functions, and connecting devices between them shall not impair their characteristics or functions.

3 Interpenetration of two different insulating media shall not occur in the oil/gas bushing between GIS and transformers or reactors, and the bushing shall be able to withstand the maximum pressure difference between normal pressure on one side and vacuum on the other side.

4 The SF_6 pipeline busbar which connects with the oil/gas bushing for transformer or reactor shall be provided with detachable breaks, and the clearance between breaks shall be able to withstand various test voltages. A separate compartment may be provided for detachable breaks and be linked with adjacent compartments during operation.

5 Insulating components should be set at the GIS enclosure connecting parts to prevent GIS enclosure induction current from being transmitted to the enclosure of transformer or reactor through the SF_6 pipeline busbar. The insulating components shall be able to withstand the maximum induction voltage on SF_6 pipeline busbar, and shall be able to withstand 2 kV power frequency voltage for 1 min. ZnO overvoltage limiters should be provided straddling the insulating components.

6 The SF_6 pipeline busbar shall be in good contact with oil/gas bushing conductive circuit. Under the normal working condition, the temperature rise shall meet the requirements of relevant standards, and the busbar shall be easy to detach for installation, maintenance and test.

7 If GIS has not been connected with transformer or reactor when commissioned, required measures shall be taken to seal the connecting point.

8 The structural dimensions and performance parameters of the connecting parts between GIS and transformers or reactors shall be coordinated.

4.10.2 The connection between GIS and high-voltage cables shall meet the following requirements:

1 GIS connected to the cables through cable terminals shall comply with the current national standard GB/T 22381, *Cable Connections Between Gas-Insulated Metal-Enclosed Switchgear for Rated Voltages Equal to and Above 72.5 kV and Fluid-Filled and Extruded Insulation Power Cables—Fluid-Filled and Dry Type Cable-Terminations*.

2 GIS and cable terminals have independent performance and functions, and connection devices between them shall not impair their performance and functions.

3 Cable connection device and detachable break shall be set at the connection between SF_6 pipeline busbar and cable terminals, and the clearance between breaks shall be able to withstand various field test voltages. A separate compartment may be set for the connection device and be linked with adjacent compartments during operation.

4 The SF_6 pipeline busbar shall be in good contact with cable terminal conductive circuit. At the rated condition, the temperature rise of connection device shall meet the requirements of relevant standards, and the connection device shall be easy to detach for installation, maintenance and test.

5 Sealing measures shall be taken at the interface to prevent SF_6 gas from leaking from cable terminals. Plugging measures shall also be taken when cable terminals are not installed. Place of bushings for cable test is reserved near the cable connection enclosure if necessary.

6 Insulating components should be set between the GIS enclosure and the metal sheath of cables. The insulation components shall be able to withstand the maximum induced voltage under various operating conditions and shall be able to withstand 2 kV power frequency voltage for 1 min. ZnO overvoltage limiters, at least 3 groups, should be provided straddling the insulating components.

7 The structural dimensions and performance parameters of the connections between GIS and cable terminals shall be coordinated.

4.10.3 The connection between GIS and overhead line shall meet the following requirements:

1 When GIS is connected to overhead line through SF_6/air bushing, the layout of SF_6/air bushing shall comply with the current sector standard DL/T 5352, *Technical Code for Designing High Voltage Electrical Switchgear*, and the convenience of connection with overhead line shall be also considered.

2 The requirements for safe clearance, insulation level, creepage distance, mechanical strength and terminals of SF_6/air bushing shall comply with the relevant standards.

4.10.4 The connection between GIS and GIL shall meet the following requirements:

1 Taking the insulated partition connecting GIS and GIL as the interface, GIS manufacturer is responsible for the design of the insulated partition, and GIL manufacturer should be responsible for the design of conductor connection and enclosure connection between insulated partition and GIL, and sealing components.

2 Insulated partitions shall be able to withstand 2 kV power frequency voltage for 1 min.

3 Detachable breaks shall be set for the SF_6 pipeline busbar connected to GIL, and the clearance between breaks shall be able to withstand various field test voltages. A separate compartment may be set for the connection device and be able to connect with adjacent compartments during operation.

4 The materials, structural dimensions and performance parameters of the enclosures and conductors at the connection between GIS and GIL shall be coordinated.

4.10.5 The interface of GIS low-voltage cables should be the local control unit or marshalling kiosk for each bay. The connection between GIS and local control units and between local control units and the connection in the local control units should be designed by GIS manufacturer, and the connection between the local control units and other equipment should be designed by the user.

4.10.6 Other interfaces shall meet the following requirements:

1 For GIS foundation fixing, the expansion bolts, if used, should be designed by GIS manufacturer, and the embedded parts, if used, shall be designed by the user.

2 The earthing busbar of GIS and the earthing leads from GIS equipment to earthing busbar should be designed by GIS manufacturer, and the connection between earthing busbar and earthing grid shall be designed by the user.

3 The ancillary facilities for operation, patrol and maintenance of GIS should be designed by GIS manufacturer.

5 Type Selection of Components

5.1 Type Selection of Circuit Breaker

5.1.1 The circuit breaker shall be of vertical type or horizontal type with due consideration of such factors as site location and size, main electrical connection, civil structures, way of line incoming and outgoing, equipment transportation and installation conditions, manufacturing experience of the supplier, etc.

5.1.2 The circuit breaker of 330 kV and below shall adopt a single break.

5.1.3 The circuit breaker of 500 kV and above with double breaks shall adopt the same operating mechanism.

5.1.4 When the circuit breaker adopts double breaks, GIS manufacturer shall conduct electromagnetic resonance numerical calculation of the no-load busbar for circuit breaker when breaking with PT and take measures to prevent ferromagnetic resonance.

5.1.5 Whether the GIS line bay circuit breaker of 330 kV and above is provided with closing resistor shall be determined after verifying the feasibility of only using surge arrester to restrict closing and reclosing overvoltage according to project conditions and operation modes.

5.2 Type Selection of Pipeline Busbar

5.2.1 The type of pipeline busbar shall be selected according to factors such as the capability of the manufacturer, operating experience, site size, the influence degree on the power system when a three-phase fault occurs, and the influence on the overall structural layout of the GIS. The selection of pipeline busbar should meet the following requirements:

1 Three-phase pipeline busbar should be adopted for GIS of 110 kV and below.

2 Three-phase or single-phase pipeline busbar may be adopted for 220 kV and 330 kV GIS.

3 Single-phase pipeline busbar should be adopted for GIS of 500 kV and above.

5.2.2 Enclosures of single-phase and three-phase pipeline busbars shall be made of aluminum alloy.

5.3 Other Components

5.3.1 The types and combinations of disconnectors, earthing switches, current transformers, voltage transformers, surge arresters and other components shall be selected according to the main electrical connection, overall layout of the GIS, and the product characteristics of the manufacturer. The type selection of equipment shall prevent the reactance of the voltage transformer from resonating with the capacitance of other GIS equipment.

5.3.2 The equipment for expansion projects or those constructed in stages should have the same type as the existing equipment.

5.3.3 Circuit breaker, disconnector and earthing switch shall be provided with reliable and readily accessible closing and opening position indicators.

6 Layout

6.1 Location

6.1.1 The location of GIS shall be determined through comparison of alternatives according to the overall layout characteristics, to facilitate incoming and outgoing lines, shorten the connection, reduce the SF_6 pipeline busbar, elbows and the quantities of civil works, and make full use of existing structures.

6.1.2 The layout of GIS shall fully consider the installation and operation environmental conditions, convenience of inspection and observation, cost of operation and maintenance, loss due to power outages, investment and the service life of GIS, etc. Priority should be given to indoor arrangement with favorable installation and operation environment.

6.1.3 Indoor arrangement should be adopted for the following project conditions:

1 Restricted site areas.

2 Severe cold areas with the lowest temperature below −25 °C and extremely hot areas with the highest temperature above +40 °C.

3 Heavily polluted areas with air pollution level d or above.

4 Cement fog area, coastal area, heavy hail area, etc. with harsh operation environmental conditions.

5 Areas with average daily relative humidity exceeding 95 %.

6.1.4 The layout of GIS connected with overhead incoming and outgoing lines should avoid cement fog area, otherwise the creepage ratio of SF_6/air bushing shall be increased.

6.2 Layout Principle

6.2.1 The layout of GIS shall be determined according to the main electrical connection, the type of GIS, the layout of main transformer, the mode and layout of incoming and outgoing lines, the installation, operation and maintenance, etc.

6.2.2 GIS should be arranged in a three-phase group, or may be in an identical-phase group.

6.2.3 The foundation of GIS equipment in the same bay shall not cross the structural joints of civil works.

6.2.4 For the arrangement sequence of GIS busbar, when arranged horizontally, the transformer side busbar is Busbar Ⅰ and the line side busbar is Busbar Ⅱ; when arranged in two layers, the lower is Busbar Ⅰ and the upper is Busbar Ⅱ.

6.2.5 A marshalling kiosk shall be set for each bay, which may be arranged on the wall side of the main channel of circuit breaker or may be arranged in parallel with the circuit breaker, and also may be arranged on the equipment without affecting the installation and maintenance of equipment. The marshalling kiosk of GIS of 220 kV and below and with a small number of bays may be of centralized arrangement, but the relative position of marshalling kiosk of each bay to the corresponding circuit breaker and voltage transformer should be consistent.

6.2.6 SF_6/air bushing shall adopt medium-height arrangement, and the convenience of line jumper connection should be considered. SF_6/air bushing of 220 kV and above should not be horizontally arranged.

6.3 Site and Passage

6.3.1 The layout of GIS shall consider the space and passage required for installation, maintenance, lifting, inspection, on-site withstand voltage test equipment transportation and SF_6 gas recovery devices handling, and also reserve the installation site.

6.3.2 GIS shall be provided with passage for transportation, installation, maintenance and inspection. The main passage should be close to the circuit breaker side, and the passage width shall not be less than 2 m. The passage width for GIS of 330 kV and above shall not be less than 2.5 m. The inspection passage width on the other side should not be less than 1.0 m, and shall not be less than 0.8 m for confined space.

6.3.3 The lifting and handling space required for the largest transport unit during installation and maintenance shall be checked.

6.3.4 Mobile or fixed platforms and ladders should be set when the GIS parts that need to be operated, inspected and maintained frequently are high above the ground.

6.3.5 For GIS with a large number of bays, a passage of not less than 1.0 m laterally should be set at the appropriate location.

6.4 Phase Sequence

6.4.1 The phase sequence of different circuits of GIS should be consistent. Generally, the phase sequence is A, B and C from left to right, from far to near and from top to bottom, when facing the outgoing current direction. The phases of equipment and busbar shall be marked yellow (A), green (B) and red (C).

6.4.2 The phase sequence of expansion or renovation project should be consistent with that of the existing installation.

6.5 Layout of Site Withstand Voltage Test Equipment

6.5.1 The layout of withstand voltage test equipment on site shall be such that the connection line between the test equipment and GIS is the shortest. The load, transportation and electrical clearance of test equipment shall be considered for the site.

6.5.2 When GIS adopts SF_6/air bushing for the connection with overhead line, the test equipment should be close to the bushing, taking SF_6/air bushing as connection point. When it is difficult to directly connect with SF_6/air bushing restricted by site condition, the connection point may be set on the first span of overhead incoming and outgoing lines, and the capacity of test equipment shall be checked.

6.5.3 When GIS is connected with cable or long GIL, busbar terminal or the connection of voltage transformer should be selected as the connection point of withstand voltage test and provided with special test bushing, and the arrangement space and electrical clearance for the test equipment shall be ensured on site.

6.6 Extension

6.6.1 The design of GIS shall be fully considerd with the planned extension of the project, if any, and reserve the space required for the extension.

6.6.2 The extension interface shall be preferentially connected with pipeline busbar and should not be directly connected with the compartment containing operable switchgears, such as circuit breaker and disconnector.

6.6.3 If the user requires extension to connect with another GIS product, the manufacturer shall provide the data required for the interface design at expansion stage in the form of drawings.

6.7 Layout of Low-Voltage Cables

6.7.1 The cables for control, measurement, signal, protection and power of GIS should be arranged by bay, and be led to the marshalling kiosk of this bay through cable channel boxes, and then led to the central control room or other places.

6.7.2 The control, measurement, signal and protection cables in bay shall be of shielded type, and shall be led to the marshalling kiosk separately from the power cables.

6.7.3 The signal, control, protection and power cables should be combed separately at the gathering place, and enter the marshalling kiosk from different directions. The convenience of cable installation and maintenance shall be considered when laying, and the requirements of cable bending radius shall be met.

6.8 Supporting Structure and Foundation

6.8.1 GIS support should be of adjustable type.

6.8.2 The expansion bolts should be adopted for fixing GIS support to the foundation.

6.9 Crane

6.9.1 Overhead crane shall be arranged in GIS room of 220 kV and above, and should be dual-speed operation in three directions. For the GIS room of 110 kV and below, overhead crane or hooks may be arranged. The lifting capacity and height of the overhead crane or hooks shall meet the requirements of lifting the largest equipment.

6.9.2 The overhead crane in GIS room should be operated on the ground.

6.10 Arrangement of Auxiliary Equipment

6.10.1 Depending on the operation mode, GIS may be provided with a dedicated maintenance room, spare parts warehouse and SF_6 gas storage site.

6.10.2 The dedicated maintenance room shall consider the needs for removing circuit breaker arc extinguishing chamber and replacing contacts, be dustproof and provided with simple air conditioner and lifting equipment. The special monitoring equipment and the spare parts with higher environmental requirements may be placed in the maintenance room.

6.10.3 A separate spare parts warehouse may be set up for large projects, and special tools and spare parts may be stored in the spare parts warehouse.

6.10.4 SF_6 gas shall be stored in the special storage room with ventilation conditions and kept away from heat sources.

6.11 Additional Requirements for Outdoor GIS

6.11.1 The influence of environmental and meteorological conditions shall be considered for outdoor GIS, and the requirements for sealing and corrosion resistance shall be put forward.

6.11.2 The outdoor GIS shall meet the following requirements:

1 Anti-aging and corrosion-resistant measures shall be taken for the exposed part of GIS flange sealing ring.

2 The installation position of density relay should not be too high, and the influence of solar radiation, refraction, reflection, and dust accumulation on observation shall be avoided.

3 Rain covers shall be provided for the density relay, instruments and apparatus, cable junction box and charging/discharging interface.

4 The exposed rotary parts of operating mechanism shall be protected against rain, snow, ice and hail.

5 The cables of auxiliary circuit shall be protected from sunlight exposure, and proper measures shall be taken to prevent the cable pipe from aging and cracking.

6 The waterproof, moisture-proof and ventilation measures shall be taken for cable channel boxes.

7 The protection level of outdoor marshalling kiosk shall not be lower than IP54. The kiosk body shall be ventilated and provided with functions of anticorrosion, rainproof, moisture proof, dustproof and preventing small animals from entering.

7 Environment Protection

7.1 General Requirements

7.1.1 The connections and associated holes between GIS and adjacent offices, equipment rooms, cable galleries, cable trenches, and stairways shall be clogged or necessary measures shall be taken to prevent the leaked SF_6 from entering these places.

7.1.2 The locations where GIS crosses roof and wall shall be provided with sealing device against rainwater leakage.

7.2 Environment Protection of GIS Room

7.2.1 The oxygen content of the air in the GIS room shall be more than 19.5 %, and the SF_6 gas concentration shall be less than 1000 μL/L (6 g/m^3).

7.2.2 The GIS room shall be provided with SF_6 leakage monitoring device to detect SF_6 gas concentration or oxygen content in the air. For GIS with fewer bays and voltage level of 110 kV and below, mobile leakage monitoring device may be used. For GIS of 220 kV and above, fixed monitoring device shall be used and the alarm signal shall be sent to the central control room. When the SF_6 gas concentration or oxygen content in the air does not meet the requirements in Article 7.2.1 of this code, the monitoring device shall send out alarm signal and automatically start the ventilation system. The ventilation system shall remain operating for 15 min after the alarm is cleared.

7.2.3 The ventilation design of the GIS room shall comply with the current sector standard NB/T 35040, *Design Code for Heating Ventilation and Air Conditioning of Hydropower Plants*.

8 Earthing

8.1 General Requirements

8.1.1 To ensure personal and equipment safety, the main circuit, auxiliary circuit, equipment support frame and all metal components of the GIS shall be reliably earthed.

8.1.2 For single-phase type GIS, electrically continuous enclosure and multi-point earthing shall be adopted. For three-phase type GIS, multi-point earthing should be adopted.

8.1.3 The earthing design of GIS enclosure shall ensure that the induction voltages of equipment enclosure, frame and easily-touched parts do not exceed 24 V under the normal operation condition. The induction voltage under the fault condition shall comply with the current national standard GB/T 50065, *Code for Design of AC Electrical Installations Earthing*.

8.1.4 GIS shall be provided with an exposed earthing busbar that runs through all GIS bays, and large GIS shall be provided with an exposed looped earthing busbar.

8.1.5 The earthing leads of GIS enclosure shall be directly connected with earthing busbar, and the earthing points of devices shall not be connected in series to earthing busbar. There shall be at least 4 connecting wires between earthing busbar and earthing grid and the connection points between earthing busbar and earthing grid should be added for long earthing busbar.

8.1.6 The earthing busbar and GIS enclosure earthing leads should be made of copper. When the earthing leads are made of different material from enclosure or earthing busbar, measures against electrical corrosion shall be taken at the contact points.

8.1.7 Except for the single-point earthing, the connections with other equipment and the external current transformer, the GIS enclosure shall have a solid electrical connection, and jumper wires shall be provided at the buffer junction and the insulating partition. The cross-section of the jumper wires shall meet the requirements for long-term maximum current passing through the enclosure. When the GIS enclosure is made of aluminum alloy, aluminum bars should be used for jumper wires; when it is a steel enclosure, copper bars should be used for jumper wires.

8.1.8 The GIS enclosure should be connected to the support frame through an insulator, or may be directly connected. The connection method shall meet the following requirements:

1 When the support frame supports the enclosure using insulating blocks, the insulating blocks shall be able to withstand a power frequency withstand voltage of 2 kV for 1 min, and have sufficient support strength and high temperature aging resistance.

2 When directly supporting the enclosure, the support frame shall be earthed and ensure that its temperature rise does not exceed 30 K under various operating conditions.

8.1.9 The cross-section of the GIS enclosure earthing lead may be selected based on 70 % of the maximum single-phase short-circuit current for multi-point earthing and shall be selected based on the maximum single-phase short-circuit current for single-point earthing. The cross-sections of the earthing busbar and the connecting wires with the earthing grid shall be selected based on 70 % of the maximum single-phase short-circuit current.

8.1.10 For the time required for thermal stability verification of the earthing, the backup protection action time should be selected for the switchgear installation of 220 kV and below, and the failure protection action time should be selected for the extra-high voltage switchgear installation of 330 kV and above.

8.1.11 The GIS earthing grid shall be provided with appropriate temporary earthing points and distinct signs to facilitate installation, maintenance and test.

8.2 Earthing of Single-Phase GIS

8.2.1 The three-phase enclosure of GIS shall be provided with short-circuit wires and meet the requirements of Article 8.1.3 of this code.

8.2.2 When GIS is directly supported by the support frame, a three-phase enclosure short-circuit wire shall be set at the supporting frame, and an earthing lead shall be set at the enclosure short-circuit wire and led to the nearest earthing busbar.

8.2.3 The end of the connection between the GIS and other equipment shall be provided with an earthed three-phase enclosure short-circuit wire.

8.2.4 The short-circuit wire of the three-phase enclosure shall have such a cross-section that it can withstand the long-term maximum induced current and the maximum short-circuit current, should be made of the same material as the GIS enclosure, and should adopt copper bars when the GIS enclosure is made of steel.

9 Requirements for Civil Design

9.1 Requirements of Indoor GIS for Civil Works

9.1.1 The GIS room shall be clean and dry. The floor, ceiling, walls, doors and windows of the GIS room shall meet the following requirements:

1. Cement mortar floor with epoxy paint, terrazzo floor or plastic floor should be adopted.
2. The ceiling and walls of the room should be painted.
3. The door shall be a fire-resistant door that opens outwards, with a spring lock, and no latch shall be used. When a door is set between adjacent switchgear installation rooms, it shall be able to open in two directions.
4. Windows should not open towards other rooms. The rooms located in severe environmental conditions and seriously polluted areas should be provided with double-pane or double-glazing windows, and measures shall be taken to prevent rain, snow, small animals, sandy wind, and dust from entering.

9.1.2 Exits shall be arranged at both ends of the GIS room. The aisles and accesses in the room shall not be obstructed.

9.1.3 Moisture-proof measures shall be taken for underground GIS room, which shall be free of groundwater leakage. Leakage-proof partition walls and waterproof roofs shall be used if necessary.

9.1.4 The civil construction error of GIS room shall meet the following requirements:

1. The displacement on both sides of the concrete foundation slit line shall not exceed ±10 mm in horizontal and longitudinal directions, and ±5 mm in vertical direction.
2. The civil construction error accumulated to the installation surface of the GIS shall be ±8 mm horizontally and ±8 mm vertically.
3. The unevenness of the floor surface in the GIS room shall not exceed ±10 mm.
4. The uneven settlement of the foundation shall not exceed 10 mm during the GIS operation.

9.1.5 For the foundation load of GIS installation, the static load of equipment, dynamic load during circuit breaker operation, earthquake load, and load of on-site withstand voltage test equipment, etc. shall be considered.

9.2 Requirements of Outdoor GIS for Civil Works

9.2.1 The unevenness of the ground surface in the area of GIS installation, the uneven settlement of the foundation during GIS operation, and the displacement on both sides of the concrete foundation slit line shall comply with Article 9.1.4 of this code.

9.2.2 The load design of GIS foundation shall comply with Article 9.1.5 of this code.

9.2.3 Concrete foundation shall be adopted for GIS installation. Concrete accesses for installation, operation and maintenance and outdoor equipment area should be filled with gravel with the particle size of 20 mm to 30 mm. Other areas should be greened.

9.2.4 Drainage measures shall be considered for GIS installation area, and there shall be no ponding on the floor surface.

10 Allocation of Special Tools and Instruments

10.0.1 GIS installation should be provided with a full set of special tools for handling, installation, maintenance, and lifting of all components and equipment.

10.0.2 Considering the operation and management mode, GIS should be provided with the following tools and monitoring instruments.

1 SF_6 gas recovery device.

2 SF_6 gas replenishing trolley.

3 SF_6 gas humidity detector.

4 SF_6 gas leakage detector.

5 SF_6 gas analyzer.

6 Monitoring device for SF_6 gas concentration or oxygen content in the GIS room.

7 Breaking and making speed measuring instrument for circuit breaker, disconnector and rapid earthing switch.

11 Site Test

11.1 Test Items

11.1.1 The site test items of each GIS component shall be in accordance with the relevant standards.

11.1.2 The site test items of GIS shall comply with the current standards of China DL/T 618, *Guide for Hand-Over Test of Gas-Insulated Metal-Enclosed Switchgear on Site*; GB 50150, *Electric Equipment Installation Engineering-Standard for Hand-Over Test of Electric Equipment*; and GIS procurement contract documents. Site test and check shall include at least the following items:

1 Main circuit resistance measurement.

2 Acceptance of new SF_6 gas.

3 SF_6 gas sealing test.

4 SF_6 gas humidity measurement.

5 Withstand voltage test on the main circuit.

6 Partial discharge test.

7 Insulation test on auxiliary circuits.

8 Inspections of interlocking and locking.

9 Calibration of SF_6 gas density relay and pressure meter.

10 Mechanical operation and mechanical characteristics test.

11 Inspection and verification.

11.1.3 Test items of GIS cooperating with system debugging shall be specified in GIS procurement contract documents.

11.1.4 If a circuit breaker or disconnector needs to be disassembled due to its collision in transportation or damage in installation, consultation shall be made with the manufacturer for carrying out the withstand voltage test between breaks after re-installation.

11.1.5 Partial discharge test shall be carried out when site conditions permit. Partial discharge test shall comply with the current sector standard DL/T 618, *Guide for Hand-Over Test of Gas-Insulated Metal-Enclosed Switchgear on Site*. Partial discharge test should be carried out on the same specimen after its AC withstand voltage test, or may be carried out in combination with the AC withstand voltage test.

11.2 Withstand Voltage Test on the Main Circuit

11.2.1 On-site withstand voltage test shall be carried out for the new GIS, and for the bays of the expanded GIS.

11.2.2 On-site withstand voltage test should be carried out on the connection section between the expanded bay and the existing GIS. The test shall comply with the current national standard GB/T 7674, *Gas-Insulated Metal-Enclosed Switchgear for Rated Voltages of 72.5 kV and Above*. During the test, repeated test of the existing GIS shall be restricted or avoided.

11.2.3 During the on-site withstand voltage test, GIS shall be isolated from the high voltage cable, overhead line, long GIL, power transformer, high voltage reactor, voltage transformer and surge arrester.

11.2.4 For GIS with many bays, the capacity of test equipment and the feasibility of batch test shall be verified. Repeated test of equipment components shall be reduced.

11.2.5 The test procedure, test voltage, voltage waveform, application of test voltage, and evaluation of the test shall comply with the current standards of China DL/T 618, *Guide for Hand-Over Test of Gas-Insulated Metal-Enclosed Switchgear on Site*; GB/T 7674, *Gas-Insulated Metal-Enclosed Switchgear for Rated Voltages of 72.5 kV and Above*; and DL/T 555, *Guide for Withstand Voltage and Insulated Test of Gas-Insulated Metal-Enclosure Switchgear on Site*.

Explanation of Wording in This Code

1 Words used for different degrees of strictness are explained as follows in order to mark the differences in executing the requirements in this code.

1) Words denoting a very strict or mandatory requirement:

"Must" is used for affirmation; "must not" for negation.

2) Words denoting a strict requirement under normal conditions:

"Shall" is used for affirmation; "shall not" for negation.

3) Words denoting a permission of a slight choice or an indication of the most suitable choice when conditions permit:

"Should" is used for affirmation; "should not" for negation.

4) "May" is used to express the option available, sometimes with the conditional permit.

2 "Shall meet the requirements of..." or "shall comply with..." is used in this code to indicate that it is necessary to comply with the requirements stipulated in other relative standards and codes.

List of Quoted Standards

GB/T 156, *Standard Voltages*

GB/T 762, *Standard Current Ratings*

GB 3096, *Environmental Quality Standard for Noise*

GB/T 4208, *Degrees of Protection Provided by Enclosure (IP Code)*

GB/T 7674, *Gas-Insulated Metal-Enclosed Switchgear for Rated Voltages of 72.5 kV and Above*

GB/T 8905, *The Guide for Management and Measuring* SF_6 *Gas in Electrical Equipment*

GB/T 11022, *Common Specifications for High-Voltage Switchgear and Controlgear Standards*

GB/T 12022, *Industrial Sulfur Hexafluoride*

GB/T 13540, *Seismic Qualification for High-Voltage Switchgear and Controlgear*

GB/T 20138, *Degrees of Protection Provided by Enclosures for Electrical Equipment Against External Mechanical Impacts (IK Code)*

GB/T 22381, *Cable Connections Between Gas-Insulated Metal-Enclosed Switchgear for Rated Voltages Equal to and Above 72.5 kV and Fluid-Filled and Extruded Insulation Power Cables—Fluid-Filled and Dry Type Cable-Terminations*

GB/T 22382, *Direct Connection Between Power Transformers and Gas-Insulated Metal-Enclosed Switchgear for Rated Voltages of 72.5 kV and Above*

GB/T 26218.1, *Selection and Dimensioning of High-Voltage Insulators Intended for Use in Polluted Conditions—Part 1: Definitions, Information and General Principles*

GB/T 28819, *Aluminum Alloy Enclosures for Gas-Filled High-Voltage Switchgear*

GB/T 50064, *Code for Design of Overvoltage Protection and Insulation Coordination for AC Electrical Installations*

GB/T 50065, *Code for Design of AC Electrical Installations Earthing*

GB 50150, *Electric Equipment Installation Engineering-Standard for Hand-Over Test of Electric Equipment*

GB 50260, *Code for Seismic Design of Electrical Installations*

NB/T 35040, *Design Code for Heating Ventilation and Air Conditioning of Hydropower Plants*

DL/T 555, *Guide for Withstand Voltage and Insulated Test of Gas-Insulated Metal-Enclosure Switchgear on Site*

DL/T 593, *Common Specifications for High-Voltage Switchgear and Controlgear Standards*

DL/T 618, *Guide for Hand-Over Test of Gas-Insulated Metal-Enclosed Switchgear on Site*

DL/T 5352, *Technical Code for Designing High Voltage Electrical Switchgear*

DL/T 5429, *Technical Code of Design for the Electric Power System*